ICS 27.010
F 13
Record Number: 64304-2018

Energy Sector Standard of the People's Republic of China

NB/T 34061-2018

Technical specification for storage and transportation of densified biofuel for heating boiler

生物质锅炉供热成型燃料产品贮运技术规范

(English Translation)

Issue date: 2018-04-03　　　　　　　　Implementation date: 2018-07-01

Issued by　National Energy Administration of the People's Republic of China

Energy Sector Standard of the People's Republic of China

NB/T 34061-2018

Technical specification for storage and transportation of densified biofuel for heating boiler

生物质锅炉供热成型燃料产品贮运技术规范

(English Translation)

China Water & Power Press

中国水利水电出版社

Beijing 2024

All rights reserved. No part of this publication may be reproduced, stored in a retrieval system, or transmitted in any form or by any means—electronic, mechanical, photocopying, recording or otherwise, without prior written permission of the publisher.

图书在版编目（CIP）数据

生物质锅炉供热成型燃料产品贮运技术规范 : NB/T 34061-2018 = Technical specification for storage and transportation of densified biofuel for heating boiler (NB/T 34061-2018) : 英文 / 国家能源局发布. -- 北京 : 中国水利水电出版社, 2024.6. ISBN 978-7-5226-2593-5

Ⅰ. TK227.1-65

中国国家版本馆CIP数据核字第2024YB8256号

Energy Sector Standard of the People's Republic of China

中华人民共和国能源行业标准

Technical specification for storage and transportation
of densified biofuel for heating boiler

生物质锅炉供热成型燃料产品贮运技术规范

NB/T 34061-2018

(English Translation)

Issued by National Energy Administration of the People's Republic of China
国家能源局　发布
Translation organized by China Renewable Energy Engineering Institute
水电水利规划设计总院　组织翻译
Published by China Water & Power Press
中国水利水电出版社　出版发行
　　Tel: (+ 86 10) 68545888　68545874
　　sales@mwr.gov.cn
　　Account name: China Water & Power Press
　　Address: No.1, Yuyuantan Nanlu, Haidian District, Beijing 100038, China
　　http://www.waterpub.com.cn
中国水利水电出版社微机排版中心　排版
北京中献拓方科技发展有限公司　印刷
210mm×297mm　16开本　0.75印张　30千字
2024年6月第1版　2024年6月第1次印刷

Price(定价)：￥200.00

About English Translation

This English version is one of China's energy sector standard series in English. Its translation was organized by China Renewable Energy Engineering Institute authorized by National Energy Administration of the People's Republic of China in compliance with relevant procedures and stipulations. This English version was issued by National Energy Administration of the People's Republic of China in Announcement [2023] No. 8 dated December 28, 2023.

This version was translated from the Chinese Standard NB/T 34061-2018, *Technical Specification for Storage and Transportation of Densified Biofuel for Heating Boiler*, published by China Water & Power Press. The copyright is reserved by National Energy Administration of the People's Republic of China. In the event of any discrepancy in the implementation, the Chinese version shall prevail.

Many thanks go to the staff from the relevant standard development organizations and those who have provided generous assistance in the translation and review process.

For further improvement of the English version, any comments and suggestions are welcome and should be addressed to:

China Renewable Energy Engineering Institute
No. 2 Beixiaojie, Liupukang, Xicheng District, Beijing 100120, China
Website: www.creei.cn

Translating organizations:

Academy of Agricultural Planning and Engineering, MARA

China Renewable Energy Engineering Institute

Translating staff:

CONG Hongbin	YUE Lei	MENG Haibo	FENG Jing
XING Haohan	YE Bingnan	WEN Fengrui	LIU Huan

CHEN Mingsong

Review panel members:

QIAO Peng	POWERCHINA Northwest Engineering Corporation Limited
LI Zhongjie	POWERCHINA Northwest Engineering Corporation Limited
YAN Wenjun	Army Academy of Armored Forces, PLA
QIE Chunsheng	Senior English Translator
GUO Jie	POWERCHINA Beijing Engineering Corporation Limited
CHE Zhenying	IBF Technologies Co., Ltd.
LI Lijie	Academy of Agricultural Planning and Engineering, MARA

National Energy Administration of the People's Republic of China

翻译出版说明

本译本为国家能源局委托水电水利规划设计总院按照有关程序和规定，统一组织翻译的能源行业标准英文版系列译本之一。2023年12月28日，国家能源局以2023年第8号公告予以公布。

本译本是根据中国水利水电出版社出版的《生物质锅炉供热成型燃料产品贮运技术规范》NB/T 34061—2018 翻译的，著作权归国家能源局所有。在使用过程中，如出现异议，以中文版为准。

本译本在翻译和审核过程中，本标准编制单位及编制组有关成员给予了积极协助。

为不断提高本译本的质量，欢迎使用者提出意见和建议，并反馈给水电水利规划设计总院。

地址：北京市西城区六铺炕北小街2号
邮编：100120
网址：www.creei.cn

本译本翻译单位：农业农村部规划设计研究院
　　　　　　　　水电水利规划设计总院

本译本翻译人员：丛宏斌　岳　蕾　孟海波　冯　晶
　　　　　　　　邢浩翰　叶炳南　温冯睿　刘　欢
　　　　　　　　陈明松

本译本审核人员：

　乔　鹏　中国电建集团西北勘测设计研究院有限公司
　李仲杰　中国电建集团西北勘测设计研究院有限公司
　闫文军　中国人民解放军陆军装甲兵学院
　郄春生　英语高级翻译
　郭　洁　中国电建集团北京勘测设计研究院有限公司
　车振英　一百分信息技术有限公司
　李丽洁　农业农村部规划设计研究院

国家能源局

Contents

Foreword		VII
1	Scope	1
2	Normative references	1
3	Product storage	1
3.1	Technical requirements for storage facilities	1
3.2	Technical requirements for product storage	1
3.3	Storage management	2
4	Product transportation	2
5	Acceptance and measurement upon delivery	2
6	Safety and health	2

Foreword

This standard is drafted in accordance with the rules given in the GB/T 1.1-2009 *Directives for standardization—Part 1: Structure and drafting of standards*.

National Energy Administration of the People's Republic of China is in charge of the administration of this specification. China Renewable Energy Engineering Institute has proposed this specification and is responsible for its routine management, and the explanation of the specific technical contents. Comments and suggestions in the implementation of this specification should be addressed to:

China Renewable Energy Engineering Institute

No. 2 Beixiaojie, Liupukang, Xicheng District, Beijing 100120, China

Drafting organizations:

Academy of Agricultural Planning and Engineering, MARA

Beijing Yifang Sunshine Energy Technology Co., Ltd.

Chief drafting staff:

CONG Hongbin	ZHAO Lixin	MENG Haibo	HUO Lili
YAO Zonglu	YUAN Yanwen	ZHAO Kai	FENG Jing
LUO Juan	REN Yawei	LI Lijie	WANG Guan
DONG Yifang			

NB/T 34061-2018

Technical specification for storage and transportation of densified biofuel for heating boiler

1 Scope

This standard specifies the requirements for product storage, transportation, acceptance and measurement upon delivery, safety and health during storage and transportation of densified biofuel for heating boilers.

This standard is applicable to the storage and transportation of densified biofuel for heating boilers.

2 Normative references

The following referenced documents are indispensable for the application of this standard. For dated references, only the edition cited applies. For undated references, the latest edition of the referenced document (including any amendments) applies.

GB 2894, *Safety signs and guideline for the use*

GB/T 12801, *General principles for the requirements of safety and health in production process*

GB 50016, *Code for fire protection design of buildings*

NB/T 34062, *Code for design of densified biofuel projects for heating boilers*

NB/T 34065, *General rule for test methods of densified biofuel for heating boiler*

3 Product storage

3.1 Technical requirements for storage facilities

3.1.1 The location, design and construction of storage facilities shall comply with NB/T 34062 and GB 50016.

3.1.2 Storage facilities shall be equipped with communication equipment, lighting facilities and firefighting facilities.

3.1.3 The storage places shall be equipped with gas alarms, fire alarm devices and static release devices. Testing and maintenance of lightning protection earthing devices and explosion-proof devices shall meet the relevant requirements.

3.1.4 The storage places must be kept dry, and the warehouses shall have good ventilation, moisture-proofing and firefighting facilities.

3.2 Technical requirements for product storage

3.2.1 When the products are stacked outdoors, they shall be protected from rain and sunlight with the cover over and around the stacking position, and the stack bottom shall be at least 10 cm above the ground.

3.2.2 Products shall be packaged using materials with moisture-proofing ability and slight air permeability.

3.2.3 Products shall be stored separately in different areas by their types and characteristics. Retaining walls shall be established for each storage area. Fire lanes shall be reserved between different storage areas.

3.2.4 Ventilation and transportation passages shall be left between stacks.

3.2.5 The "first-in-first-out" principle shall be followed in the product warehouse-out. Regular inspections of product packaging intactness and appearance quality are necessary to prevent mildew and pests to ensure safety.

3.3 Storage management

3.3.1 The storage keeper shall establish a book account system.

3.3.2 Storage facilities shall be marked by the types and characteristics of products stored.

4 Product transportation

4.1 Product transportation shall comply with the relevant regulations of cargo transportation authority.

4.2 Product transportation vehicles shall be equipped with rain-proof coverings, and products shall be prevented from littering during transportation.

4.3 Isolation facilities and obvious signs shall be installed in the handling areas during transportation and transfer of products.

5 Acceptance and measurement upon delivery

5.1 Acceptance upon delivery shall comply with the supply contract between the two parties, and the supplier shall provide a product quality inspection report issued by a qualified testing agency in accordance with NB/T 34065.

5.2 The measuring devices used in product acceptance upon delivery shall be verified by the national metrology institute.

6 Safety and health

6.1 Dust masks shall be provided for handling workers.

6.2 Safety signs shall be established in the workplaces in accordance with GB 2894.

6.3 Safety protection measures, safety operation regulations and fire emergency plans shall be formulated, and protective and life-saving facilities and supplies shall be provided with reference to GB/T 12801 according to production process.

6.4 Storage keeper and transporter shall formulate emergency preparedness plans.

微信号：Waterpub-Pro 微信号：悦读水电

唯一官方微信服务平台

销售分类：可再生能源

ISBN 978-7-5226-2593-5

定价：200.00 元